D1577658

A SHORT PHILOSOPHY OF BIRDS

A SHORT PHILOSOPHY OF BIRDS

PHILIPPE J. DUBOIS
AND ÉLISE ROUSSEAU

Translated by
JENNIFER HIGGINS

Illustrated by
JOANNA LISOWIEC

ALLEN

5 7 9 10 8 6 4

WH Allen, an imprint of Ebury Publishing,
20 Vauxhall Bridge Road,
London SW1V 2SA

WH Allen is part of the Penguin Random House group of companies
whose addresses can be found at global.penguinrandomhouse.com

Penguin
Random House
UK

First published by WH Allen in 2019

www.penguin.co.uk

A CIP catalogue record for this book is available from the British Library

ISBN 9780753554142

Cover design by Luke Bird
Text design by Claire Mason

Typeset in 10.6/14.75 pt Absara OT by Jouve (UK), Milton Keynes
Printed and bound in Great Britain by Clays Ltd, Elcograf S.p.A.

Penguin Random House is committed to a sustainable future for our
business, our readers and our planet. This book is made from Forest
Stewardship Council® certified paper.

To Pierre and Anne

Contents

Introduction

A BLACKBIRD SITS PERCHED ON A WALL, eyes glittering, beak bright yellow. Watch him carefully. Doesn't he look happy to be a blackbird? Hopping about on the lawn, hunting for a worm, completely fulfilled by his own existence. If only we were as satisfied with ourselves and our lives as the blackbird is. Surely we would be much more content.

In stories and legends, birds are often the bearers of knowledge, messages or new ideas. The Belgian writer Maurice Maeterlinck uses them to represent happiness in his play *The Blue Bird*. In *The Conference of the Birds*, a volume of medieval Persian poetry, each bird symbolises a different aspect of human behaviour. In *The Wonderful Adventures of Nils*, Swedish novelist Selma Lagerlöf shows wild geese taking the young protagonist, Nils Holgersson, on a fantastical voyage of initiation that changes him forever.

Athena, the Greek goddess of wisdom, has a bird for her emblem: the little owl, with its bright, golden eyes. Storks, graceful birds much beloved by parents everywhere, were said to bring babies into people's homes. Then there's the white dove with an olive branch in its beak symbolising hope, and the agile swallow whose return heralds the arrival of spring each year.

In the twenty-first century, what lessons can birds still teach us? Through these brief ornithological reflections, we hope you'll discover how these creatures of the sky can guide us in all sorts of ways, helping us to reflect on our own lives, if only we take the time to observe theirs. As humans we may claim to be the 'masters of the world', at the top of the evolutionary tree. But many scientific, sociological and behavioural studies of birds, and their age-old role in literary and mythological symbolism, suggest that they could hold up an enlightening mirror to *Homo sapiens*. What if we took the time to think about what we can learn from their social interactions, their approach to seduction, their parenting skills or even the way they take a bath?

How do birds manage their love lives? Are they faithful or polyamorous? Serene or restless? Why are some of them incorrigible travellers and others diehard homebodies? Do they prefer to watch over their young until they are fully mature or guide them to independence as soon as possible? Why do male doves take on their fair share of household chores, while ruffs are terrible chauvinists? How do birds live from day to day, enduring the elements, the wind and the rain, waiting to see the moon rise and the stars appear in the evening sky? And is it true that they hide away to die?

Our reflections on all these questions are based on the most recent research, but also on our own intimate knowledge of birds, gained through long hours of observation on riverbanks, in tropical forests or on the windswept dunes of deserts the world over. We are convinced that there is much to be learned from the winged world. Birds, nimble and spontaneous, masters in the art of life, have much to tell us – if only we will listen.

1.
EMBRACING OUR
VULNERABILITY

The eclipse of the duck

THE LIVES OF BIRDS, much like our own lives, are shaped by all sorts of events that are like little deaths and rebirths. Moulting is one example. Shedding old feathers in order to acquire newer, more beautiful ones is a yearly process of loss and renewal, and it can be difficult. Although we may lose some hair here and there, humans don't experience these regular moulting phases. Nevertheless, there are times when we too must moult, or transform. At certain key moments in our lives – heartbreak, mourning, losing a job, moving to a new home – we must start afresh in some way, change our wardrobe, our haircut or our lifestyle. But only very rarely.

If we are to be reborn successfully, we need to understand how to let something within us die. This is what the bird does when it trades in the old plumage for new feathers gleaming with health. For the bird this is vital: it can't fly if its plumage isn't in perfect condition. And although it may be less obvious, the same is true for us: if we can't detach ourselves from the past, we can't move forwards.

For birds, the moulting period is a time of vulnerability. Some moulting birds are temporarily unable to fly, as is the case for certain species of ducks. They are said to be in a state of 'eclipse' plumage, a lovely phrase used to describe a liminality that occurs while the bird waits for the essential feathers that it has shed to regrow. The bird knows it is vulnerable and keeps a low profile, not engaging in any important activities during this time. It is patient. It waits for the renewal to occur so that it can regain all its former strength and beauty.

We should do the same sometimes.

In a society that constantly pushes us to perform, we no longer know how to 'eclipse' ourselves when we feel vulnerable, taking the time we need to re-energise and to gather our strength. When we are bereaved, we're told that 'life goes on'. After a heart-break, 'there are plenty more fish in the sea', or after a pet dies, 'well it was only an animal'. Life tries to push us forwards, as though we don't have every right to retreat into ourselves and be sad, mourning the fact that after a bereavement life isn't the same, or that a beloved animal will never come back. Life will bring other joys, other meetings, it's true, but why not accept the depth of our loss? In our modern human lives, we are rarely afforded the time necessary to recover from our sadness, to nurse our wounds and to perform the necessary transformation before we re-emerge into the world.

Should we be surprised, then, that we no longer know how to fly, having had our wings clipped so often? Not to mention the number of times we have clipped our wings ourselves?

Let's allow ourselves the time to moult. Let's put on our eclipse plumage for these important, dormant periods of our lives. We will emerge stronger and more beautiful than ever before, and light as feathers.

2.

A LESSON IN EQUALITY

Doves and the division of parenting

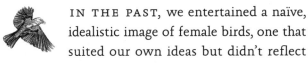IN THE PAST, we entertained a naïve, idealistic image of female birds, one that suited our own ideas but didn't reflect reality. This image portrayed the female carefully and self-sacrificingly sitting on her eggs in a nest built by the male. Meanwhile, as she sits in the nest, he prances off elsewhere, singing at the top of a tree or puffing up his chest to impress passers-by. The female, silent and drab in her dull plumage, devotes herself to her brood and feeds the chicks tirelessly while the male is already having new amorous adventures.

This caricature isn't completely false, as we see in the case of ducks. The male duck's plumage is often ornamental and very colourful, and is especially bright in spring. The female's, on the other hand, is often brown, black or white. This creates an excellent camouflage, allowing her to blend into the background of branches, reeds and grasses where she beds down into her nest and sits on her eggs. She assembles the nest herself, sometimes plucking downy feathers from her own stomach to soften the lining, like bedding. For three long weeks she sits incubating her eggs, concealed from view, never leaving her precious clutch except for a few moments to stretch her legs and find a morsel to eat. Meanwhile, the male has rejoined his gang of drakes. As soon as spring is over he begins to moult, and for a while his plumage is almost identical to that of the female. He is usually unable to fly during this vulnerable period and becomes easy prey, so much so that he will often retire to a discreet nook until moulting is over. But as for helping to bring up his young: no chance. It's the female, in her modest livery, who does all that.

As soon as the ducklings are born, she takes them to the water and doesn't leave them for an instant until they learn to fly; she is constantly on guard against the slightest hint of danger and is kept busy finding what meagre provisions she can to feed a dozen or so hungry beaks. Despite all her efforts, the initial brood of ten or twelve ducklings will shrink under the assaults of predators. By the time the young are ready to take flight, there are usually only two or three left, sometimes none. Once the young are gone the female, too, must moult quickly, because for many duck species the time of their annual migration is almost upon them. All this needs to happen in the space of a few weeks. Weakened by such hard work, the female's chances of survival are less than her mate's, so it's not surprising that in some species of ducks the males outnumber the females.

So, at one end of the spectrum sits the female duck, an incarnation of devotion, but at the other end, there are the lesser-known and rarer examples of female birds who direct operations while the male carries out his wife's orders. This is true of the large family of waders, which includes many small species (sand-pipers, snipes, plovers, stints, etc.) – all of the so-called 'beach runners' that we see on muddy, rocky or sandy stretches of coastline during their migration period in winter. Similarly, for phalaropes and the Eurasian dotterel, it is the male who does all the hard work. Phalaropes are little-known birds; they nest in the Arctic tundra and as soon as they have mated they leave to spend the rest of their existence out at sea, a remarkable life for a land bird. The dotterel (a close relative of the plover) is a beautiful species also found in the far north, migrating to North Africa and the Middle East for the winter. It is known for being very trusting of people – and indeed why should it not be, living in the middle of the tundra?

In both of these species, the female plays the role we would think of as stereotypically male. It is she who has the most brightly coloured courtship plumage, while the male is largely drab. She takes the lead in the process of seduction, carrying out the courtship display, choosing one or several males with whom she will mate and then performing all the usual seduction rituals of pursuit – mock combat with other females and parading around the male. Then, after making a token effort to help build a hole that will serve as a nest, she lays her eggs and leaves. It is up to the male to sit on the eggs for three weeks. Solitary and celibate, he must bring up the little clan. Sometimes the female returns and takes it upon herself to sound the alarm to defend the brood when danger threatens, but the male, who does not appreciate her sticking her oar in, often chases her away. The chicks are nidifugous, meaning they leave the nest as soon as they hatch out and are capable of finding food with guidance from a parent – the male parent, in this case. It must be said that although the male typically takes this role seriously, he is perhaps not as assiduous as many females of other species, and sometimes abandons the brood before they have learned to fly properly.

Some birds have conjured another solution to balance the workload: two broods – one for the male, one for the female – and each looks after their own. This is done by some small waders living in the far north. For stints and sandpipers, for example, 'married life' only lasts for a few short weeks. The female mates with the male (or sometimes with several males), lays an initial clutch in one nest and a second in another, then each partner chooses one nest and one brood, which they tend to alone. Why these single-parent families? The answer is simply that these birds spend only a short time in the distant northern regions where they breed, where the optimal window for food availability and good weather is small. By laying two clutches and leaving one to each parent they optimise their time spent at the nesting site and, most importantly, increase the chances of successful rearing.

These examples are not the norm, however. The most common practice is to take turns caring for the brood, which is, unsurprisingly, the best strategy for increasing the chicks' life expectancy. The task is easier with two. Doves are a great example of this – confirmed monogamists and also model feminists!

These birds share the tasks evenly between them; mutual assistance is key. It is the male's job to collect twigs and small branches and the female's to assemble the materials into a (rather perfunctory) nest. The same goes for sitting on the eggs: the male and female take turns, night and day. After hatching, both birds will feed the young until they take flight two weeks later. With such orderly alternation doves make a good team. This unwavering solidarity is explained by the fact that dove chicks are often targeted by predators and their nests are quite fragile and easily damaged in bad weather; the chicks may not survive, in which case the parents must begin again. A united, monogamous couple is a good solution to this problem, so good that if the first brood survives safe and sound to take flight, the adults begin the process again a few days later. If all goes well, doves can successfully raise several broods between the end of the winter and the first days of autumn.

So we see that across different species birds cover all possible strategies for dividing tasks between the male and female. And yet most people have a very simple view of avian parenting behaviour, largely due to the old naturalists (men, just in case it wasn't obvious) who waxed lyrical in their praise of the female devoted to her brood who, even when abandoned by the male, thinks nothing of sacrificing herself for the sake of her offspring – as if there's a lesson there for all of us.

If we learn one lesson from this it should be that most birds consider the sharing of tasks to be the best solution. They reached this conclusion long before us!

3.
FOLLOWING THE
RHYTHMS OF NATURE

Habits for everyday life

BIRDS ARE CREATURES OF HABIT. There is a time for eating, for drinking and for a siesta; a season for seduction, reproduction and rearing young; and for those species that undertake it, there is a period of migration. Birds' life cycles follow a very precise, ordered pattern. These habits are not mindless rituals, but actions that follow the rhythms of nature, the shifts in sunrise and sunset as the year progresses, darker or lighter nights punctuated by the phases of the moon, and constant variations in the weather and the seasons. Rain, wind, heatwaves, mist or storms – for creatures that live in the open air, life is far from predictable. Birds must constantly adapt to whatever each day brings, for better or for worse.

There are certain kinds of weather that birds don't like much, such as rain and wind. You won't see them flying in bad weather. They hide deep within a tree, hunkered down among its leaves as though they want nothing more to do with the outside world. People who keep chickens are familiar with this: when it rains or snows, the birds don't put so much as a beak out of the henhouse. They sit listlessly on their perches, sometimes for days on end, waiting glumly for the weather to change. But at the first ray of sunshine they're off outside, scratching, frolicking and enjoying the moment.

17

If our lives sometimes seem monotonous (eat, sleep, work, repeat) that's partly because, shut up in our offices in front of our computers, we don't see the seasons changing. So time passes on and passes away. Leading largely sedentary lives, indoors for hours on end, we cut ourselves off from the natural elements that are waiting there to surprise us every day, every hour, every minute. Staring at our screens until our eyes hurt, we hardly notice the little morning rain shower happening outside the window. We don't feel the sun on our skin or the wind whipping. 'What's the weather like there?' asks a far-off relative on the phone. We're almost ashamed to have barely noticed. 'Hang on, I'll look out the window ... Yes, it's a bit cloudy ...' We've hardly had time to enjoy the spring and it's already autumn. Everything passes in a blur, to the rhythm of days that all seem the same, and we don't see the grass growing, buds unfurling or grapes sweetening in the sun. The swallows, those incorrigible travellers, have gathered on the telephone wires and left for their long migration, vanishing for a whole winter – but did we look up as they disappeared from the sky? Were we struck by the absence of their twittering? No. And will we even notice when they return next spring? Probably not.

But some people work in the open air and live outdoors as animals do, and they see the swallows come and go. Just like the warbler or the wren ferreting about in the bushes, these people witness the small natural events that mark the passage of time. They know that life is not monotonous. Forest wardens, sailors, mountain guides and farmers scrutinising the clouds for signs of rain – all of these people, faced with the changing elements, must adapt, change course, plan ahead and sometimes grumble when things don't work out as they wanted. The little habits we invent for ourselves are even more enjoyable when we are faced with the instability of the weather every day. The eleven o'clock cup of coffee, regular as clockwork, might be enjoyed outside on a fine day, or provide comfort in a warm kitchen against a rainy day.

We have good habits and bad: the Sunday night film comforts and makes us feel good, whereas bad habits bore and make our lives dull. They imprison and possess, even paralyse us. But other habits have genuine benefits: they shape our lives and structure our time. They are necessary to the bird returning each year to make its nest in the same location, and just as necessary to the human being who enjoys revisiting a place full of happy memories. In a life that is already rich with change and full of the unexpected, habit can become an anchor, a beacon, a landmark. Even the greatest adventurers, who certainly can't be called creatures of habit, have their little rituals when they set off on a journey. If we retain just one lesson from birds it must surely be that when we reconnect with nature we lead richer lives filled with new, unexpected sensations.

What if we could observe the natural world around us as part of our daily routine? What if we could sharpen our senses and be a little less isolated from our environment, understanding our interactions with it a little bit more? What if we took the time to watch a bird soaring in the sky, to notice swallows twittering, to hear the blackbird's fluting song? What if we rose in the middle of the night to watch the waxing moon hovering on the horizon, full and beautiful, as an owl breaks the silence with its mysterious cry? By embracing these new, good habits, we let poetry into our lives and we say goodbye to monotony.

4.
WHATEVER HAPPENED TO OUR SENSE OF DIRECTION?

The Mongolians, the godwit and the cuckoo

IT IS JUNE 2016, somewhere deep in the Gobi Desert in Mongolia, one of the most remote and hostile places on the planet. Our expedition comprises five French explorers and six Mongolians. Our vehicles have no GPS and no mobile phone reception. There are no maps, either. They wouldn't be any use – there are not even any roads.

To guide us we have people, only people. Mongolians find their bearings from the shape of the mountains and subtle details in the natural environment that completely escape the Europeans in the group. Because, here, for tens of thousands of square miles, all the eye can see are small, undulating hills, distant mountain peaks, immense plains covered in tiny stones and innumerable little streams disappearing into gorges. Everything looks the same to the outsider and our Westerner's eyes cannot memorise a single feature, nor pick out any helpful sign to guide us. Had we been alone we would have been completely lost.

One evening, when we needed to choose between several old tracks leading off in different directions into the distance, the Mongolian leader of the expedition showed the driver which way to go without a moment's hesitation. 'When was it exactly, the last time you came here?' a member of the French cohort asked him, surprised to see him navigating so easily in this desert. 'Oh,' he replied, 'twenty years ago.' A stunned silence filled the van. He never once led us astray.

After following a barely visible path between what appeared to be two completely identical mountains, we reached the lake where we were to spend the night. The nomadic Mongolians, alongside a few other rare groups on earth, have retained an instinctive, deep-rooted sense of direction, like that possessed by migratory birds. But as for us poor Westerners, what have we done with ours?

Like Mongolians, birds don't travel with a compass, GPS or a map, because they intuitively possess all these internally. Take the bar-tailed godwit. This little wader is a close relative of the curlew and spends its life on coastal marshes or estuaries. In spring, the godwit migrates to make its nest in the Arctic. By tracking one of these godwits with a satellite tag, researchers have discovered that they are capable of covering the distance between Alaska and New Zealand – over 7,000 miles – in one go. That equates to flying for a whole week at forty-five miles per hour. Consider, too, that the godwit weighs just 250 grams. What's more, during this non-stop flight, the godwit rests by allowing only one half of its brain to fall asleep at a time – thereby enabling it to fly continuously through its sleep. Imagine if we humans could sleep this way, with one side of our brains snoozing while the other taps away on our smartphone, or carries on driving the car . . .

The cuckoo also harbours an innate sense of direction. It is born in another bird's nest with no parent cuckoos to raise it. Yet, one July evening later in its life, it will fly off to Africa, travelling by night with no previous experience of navigating. How does the cuckoo find a forest in equatorial Africa where it has never been before and, after remaining there for six months, return to the exact place where it was born? What sense, refined to an extreme degree, do these birds possess that we do not? Or rather, what have we lost?

Unlike Mongolians and migratory birds, most of us have completely forgotten our sense of direction. We no longer know how to read the landscape, the stars or the natural world. These things have become the mute decor of our environment. We have eyes but we are blind, walking and driving guided only by the directions of a robotic voice emanating from a GPS or smartphone. We have left the essential task of navigation to others – or worse still, to machines. What would we do if we were left alone in a forest or in the hills, even just thirty miles from home, and couldn't ask for directions or look at a map? How long would we spend wandering before eventually, perhaps, finding the right way? And considering all this, have we not lost the most important aspect of travel, the one that could give us real power – the basic ability to find our way, to establish the right path for ourselves? It is hardly surprising that, no longer able to orientate ourselves, many of us often feel lost in our lives. We claim to know everything, to be the masters of modern life, but in nature, even 'civilised' nature, we are as vulnerable as fledglings.

There are a few, increasingly rare, groups of people in the world who can, like the Mongolians, find their way unaided in the middle of a desert, field or deep forest. It is true that when it comes to navigation, human brains are less sophisticated than those of the bar-tailed godwit or the cuckoo, but we do possess the same tools of the natural world around us, and the capacity to navigate using the stars or even the sun's rays.

Have we really lost this vital sense? If we needed to, could human beings regain this ancient instinct over the course of a few days, or weeks, or generations? How can we know? Today we travel for pleasure without thinking about the practicalities, except for finding the cheapest plane ticket. We do away with distance and time, to the detriment of a faculty that we have now almost completely lost.

What does the bar-tailed godwit consider as it flies for seven days over the Pacific Ocean, between sky and sea? How does it pass the time? There comes a day when it slows down, flies at a lower altitude and begins to make out the traceries of marshes and rivers in the High Arctic, where everything looks the same to most human eyes. It finally lands, exhausted, in the exact spot where it built its nest the previous year.

Today, we travel much faster than godwits or cuckoos can fly – but is this really progress?

Who knows why the cuckoo and all the other migratory birds do what they do? But perhaps *how* they do it matters more than *why*. When we are preparing for our own summer 'migrations' we consult guides, maps and the Internet, and then as we travel we use radio updates, satnav, signposts . . . a whole arsenal of tools to keep us on the right track. Migratory birds have only their determination, the sea and the landscape spread out beneath their wings, below the stars, sun and moon. And yet, providing they don't perish on their long journeys, they almost always arrive in the right place.

Animal migration – and especially bird migration – is still largely a mystery to us. But we know one thing for sure. These creatures are in complete possession of their powers as they harness their skills to the fullest degree. Nomadic Mongolians still possess these

5.
WHAT IS A FAMILY, ANYWAY?

The moral of the cuckoo and the goose

WE MAY THINK WE KNOW what a family is – it seems obvious – but in fact it's far from simple. The notion of family varies widely among birds, from the cuckoo that abandons its brood before they are even hatched, to geese and cranes that maintain family ties long after the young have flown the nest.

So what is a family? Is it an innate structure indivisible from reproduction, or a result of evolution? Amoebae don't have families. It's mainly larger, more intricate animals – especially mammals and birds – that operate in family structures, all with varying degrees of complexity.

Human families are, in fact, difficult to define. Our definition of family is culturally specific and people don't always agree on what it constitutes. For some, family can only ever consist of the nuclear norm (a man, a woman, children), while others accept alternative possibilities (single-parent, blended or same-sex parental families, for example). These differences of opinion inspire endless and sometimes hostile debates. Those who only accept very traditional notions of family often use the idea of 'nature' or 'biology' to justify their beliefs, but they are forgetting that nature is very flexible on this matter. Nature's definition of a family goes something like this: an association of individuals which allow the young to be raised successfully.

Nothing more, nothing less. The important thing is not who the individuals are, but that the young grow up and achieve independence.

The family is where much of a young person's education takes place, and this education requires the parents to perform the roles of caring, teaching and protecting. At least, this is the ideal situation; in reality, some parents are unreliable and others over-protective; there are absent fathers and neglectful mothers. The same goes for birds.

The cuckoo is, in our moralising human eyes, an example of neglectful parenting. Cuckoo pairs meet only to mate, and then separate. The female lays her egg in another bird's nest and disappears as soon as this is done. The poor hoodwinked host parents must now feed a chick of a different species that is often four times their size and will immediately eliminate its foster brothers and sisters by methodically kicking each one out of the nest. And yet the host parents, driven by their reproductive instinct, play their role of parents to the cuckoo chick as best they can.

There are other cases of adoption besides the cuckoo in the bird world, many of them voluntary, either within the same species or between species. Humans are not the only ones capable of adoption.

In some bird species the male is there simply to

procreate. When the deed is done he leaves the mother to incubate and raise the young. Ducks are the champions of this method. Among some small waders it is the female who, once the eggs are laid, goes off and leaves her partner in charge of the family, as we have seen. And then there are some birds whose parenting techniques are similar to many human societies: cranes, swans, geese and storks all share the incubation and raising of their young between both parents. Even among these species, though, there are variations in the way that the young are raised.

While storks see no more of their offspring once they are old enough to be independent, this is not true of many other species, where the young may stay with their parents for several weeks after they learn to fly. Geese, for whom social bonds are very strong, take this even further. The family maintains ties between parents and young for the whole of their first winter, a long time in animal terms. For these migratory birds the continued bond is vital, helping the young geese get to know migration routes and wintering grounds. The tie with their parents gradually weakens during the winter and when the warm weather comes, the parents finally reject the young.

Community life does exist among birds, with some working as a group to raise their chicks. Pink flamingos live in colonies where parents share the task of raising the young. A few days after hatching, the chicks gather together in a 'nursery' that is easier for the adults to supervise. The parents then visit the crèche to feed their own offspring, whom they recognise.

Bird families, then, come in all shapes and sizes, from single-parent units to shared parenting groups. But it is striking to note that for birds, it is the *parents* who break the tie with their young, letting their off-spring know – sometimes in quite direct terms – that it is time they fend for themselves. This psychological weaning, which often happens after the physical wean-ing, can sometimes even become aggressive if the adolescent bird objects too strongly. This can often be observed in farmyards: a young hen whose mother has rejected it will often be forcibly pushed away with a few fierce pecks.

There is no species, either bird or mammal, in which the young stay with their parents for as long as human children do. Has the lengthy dependence of human children become pathological? Adolescence, that period of tension between parents and children, is the time when, naturally, the human parent feels the need to 'break the tie' with the child and the child wants to become independent. No bird wants to risk pecking its offspring once they are large and sexually mature, and the irritation human parents can feel towards their teenagers is surely linked to an ancient and well-founded instinct. But human parents are encouraged to be patient with their teenagers and teenagers are encouraged to stifle their desire to leave in order to finish school: both sides must therefore suppress their instincts and postpone the natural process of emancipation for a few more years, a situation that does not exist, of course, in the animal world.

Our complex society and all the knowledge needed to live in it does not make it easy for our young to take flight. And, of course, animals don't count on their young to look after them when they're old! Chimpanzees live for about forty or fifty years and their young are independent at the age of five or six. A young chimpanzee therefore spends about ten to fifteen per cent of its life dependent on its parents; for greylag geese this figure is more like six to eight per cent of their average lifespan; for *Homo sapiens* it is more like twenty-five per cent, and sometimes much more.

Our modern human familial lives are now full of social activities and educational institutions that have transformed the idea of family into a complicated social construction. While developing us as social beings, these changes have also created artificial barriers between ourselves and our physiological and environmental realities.

Family is also the fixed point around which future generations are conceived; its influence spreads far

6.
TRUE COURAGE

The eagle and the robin

HUMAN BEINGS LIKE TO VIEW the living things around them in an anthropomorphic light. Sentimentally, we insist on giving flowers a language: the rose says one thing, the cornflower another. When it comes to animals, especially larger, more complex animals, we tend to endow them with attitudes and behaviours that are purely the result of our own subjective interpretation.

Let's take the example of the eagle, which is a symbol of strength and power, the lion of the bird world. Countless countries and political parties have chosen it as their emblem. With its easy mastery of the air, the eagle is majestic in flight, like most birds of prey. Its yellow eyes bestow a hard, cold, even virile gaze and, thanks to its unrivalled eyesight (we say 'eagle-eyed' for a reason!) it can detect the slightest movement from several hundred metres above the ground. But this tough image shatters as soon as the eagle opens its beak and rasps a series of inelegant croaks.

The eagle pursues the path of least resistance with dedication. You'll never catch one mimicking a falcon's dazzling acrobatic flights in pursuit of its prey. It's much more likely to glide, carried on the air, and use its strength and its weapons alone – its beak and formidable talons – to capture its prey.

However, when it comes to defending its territory

the eagle is not the most courageous beast. It's not cowardly, but its reputation for bravery is surprising. If political leaders were hoping to find a truly warlike bird, or a symbol of real courage to adorn their flags, they would have done better to choose the robin. For all its sweet, garden-loving appearance, here is a bird that is a real brawler. The robin is a little ball of feathers that never backs down and can't stand it when a neighbour invades its personal space, even though it likes nothing better than to encroach on someone else's territory itself. It is a warmonger, proclaiming as much through its deceptively melodious and melancholy song. It is such a fierce defender of its territory that it will fight with its own reflection in a window or car wing mirror. Still, a robin, at just fourteen centimetres tall, works a little less well on a flag or a sword than an eagle with a two-metre wingspan in all its regal glory.

When it comes to bravery, we should also consider the case of the cockerel, France's national emblem. In the farmyard, the gander (the male goose) is far superior to the cockerel at pecking your shins in defence of his mate and young. And no cockerel or hen comes close to the gander in terms of vigilance: he is as good as a guard dog. The cockerel, on the other hand, flees when he is frightened, emitting disapproving clucks in a display that couldn't be further from the swagger he adopts for parading around the chicken coop. But a country that chose the gander as its symbol would be represented by a rather chubby, ungainly bird with a neatly ordered family life – one male and one female united for life. The cockerel, on the other hand, exudes self-assurance, with his shiny plumage, haughty gaze and lascivious lifestyle. No wonder the French people identified him as their emblem! The Christian world also adopted the cockerel as a symbol – of the transition from night to day, from darkness to light, although his song is by no means as beautiful as that of other dawn visitors, such as the blackbird or thrush.

The cockerel is an endearing animal by many accounts, and far from stupid, but does that really justify its status as a national emblem? For the French, the Romans probably had something to do with it. The whole concept for the country's national symbol appears to stem from a play on words between the Latin word for cockerel, *gallus*, and the word for Gaul, *Gallia* (although the Gauls, by the way, had no particular affinity with the bird in question, except when it came to eating it). The jokes weren't long in coming, and some are very well-worn by now. Seneca once observed that '*gallus in sterquilinio suo plurimum potest*', which translates as 'the cockerel is king on his own dung heap', or, more figuratively, 'even the beggar is master of his own house'. The phrase has since been rather mischievously misinterpreted to imply that the French chose the cockerel because he's the only animal that sings even when he's in the shit.

It would seem, then, that in the past powerful men liked to claim aggressive, rowdy or powerful species as their ornithological symbols. Perhaps if women had been in charge they would have chosen very differently. Terns, for example: adventurous travellers, unfailingly elegant, who always come to one another's aid.

All too often, we humans mistake strength for courage, power for bravery. We are too caught up with appearances. Watching birds teaches us that small species can successfully stand up to larger ones. By puffing up its feathers, flapping its wings and emitting loud cries, a small bird can see off a far larger adversary. A tern, for example, will set off in violent pursuit of a seagull that has flown over the colony in search of chicks. The gull will be forcibly seen off by the tern, which is not afraid to give it a few good pecks. Some birds have been observed to land on the backs of larger birds of prey in flight and attack their heads with sharp little beaks. Furthermore, birds with dull plumage are often the most effective in defending their territory or young. Beautiful plumage does not make a thick skin. Males with coloured plumage are more likely to run away or hide than to confront the danger. Meanwhile, the beguiling little robin is a valiant foe, capable of holding its own among its fellow birds, seeing off intruders and even standing up to cats on occasion. The same goes for the little owl, which, despite being one of the smallest owls, doesn't balk at attacking predators. These birds rely on their inner fearlessness and determination, not their appearance. Did the rulers of long ago really take the time to observe their birds carefully before choosing which one to adopt as their national symbol?

7.
WHAT'S LOVE GOT TO DO WITH IT?

The tenderness of the dove

TWO DOVES ON A TELEPHONE wire in spring, preening one another for hours on end. They stroke each other tenderly with their beaks, around the eyes, along the neck and over the head, completely absorbed in this intimate moment, eyes closed with pleasure. They feel good, the pair, huddled together on their wire, warming themselves in the sun and exchanging a thousand caresses. Nothing seems to trouble them. They're happy. They're in love.

Aren't they just like all lovers? Like the ones who exchange kisses on parks benches, arms entwined, hearts on fire, immersed in their beloved's gaze, dazzled, drunk with happiness, certain that all this will last forever?

Birds are often chosen above all other animals as symbols of love. Rabbits and crocodiles are less romantic, and they do say that love gives us wings. Images of white doves, symbols of conjugal happiness, often appear at weddings, and the nightingale is the lovers' companion, filling the warm summer dusk with its lyrical, full-throated song. And then there are the little parrots that are known as lovebirds because they're so devoted and tender towards their partners. It's hard to think of them without being reminded of those old married couples who are still so in love, several

47

decades on, that when one dies the other follows just a few days later.

When it comes down to it, what does it mean to love? Do birds, like us, fall in love at first sight? Can they develop friendships? Might they say of their companionship, as Montaigne said of his friend La Boétie, 'Because it was him; because it was I'?

While some species of bird practise the 'every man for himself' approach, others form strong bonds with those around them. This is the case for greylag geese, who become deeply attached to their partners and their young during the first year of their lives, guiding them when they are old enough to undertake their first migration. Similarly, the mother hen's care for her chicks has made her a symbol of maternal love.

Love that develops into bonds of friendship also appears to exist among birds, given that some cannot do without their comrades and live in large groups, such as bearded reedlings or parrotbills. Birds raised in captivity are extremely affectionate towards the humans raising them, and even towards animals of other species. Ties of friendly companionship can therefore develop outside the realms of reproduction. Loving, after all, describes the forging of a link or relationship with one or several individuals that brings benefits, and often pleasure, to all. It also leads to unhappiness when, for whatever reason, that link is broken.

The notion of love in its broadest sense has occupied philosophers since time immemorial, and still gives us food for thought. We don't all define love in the same way. What is true love? Now there's a question! There are endless different forms of attachment. Love between lovers, filial love, brotherly love, friendship . . . And it's likely that some birds experience most of these states. They can be overcome with emotional and sexual passion just as they can experience moments of tenderness when they preen one another's plumage or look after their young.

Birds probably don't experience all the nuances and subtleties of human love (just as, vice versa, they don't experience the torments of hatred that human beings can undergo). Nevertheless, when we watch a pair of doves it's clear that they feel tenderness, respect, physical attraction, goodwill and the desire to help one another. Similarly, when a cockerel finds some delicious grains of corn to eat, he calls his hens to share his booty with him and appears to be full of pride, just like a suitor bringing croissants to his beloved in the morning.

Perhaps birds are showing us what love is. This blend of tenderness and respect, of attraction, goodwill and sensitivity, the desire to do good things for the one we love, not to hurt or harm them, to offer them little presents or good things to eat, to be in sympathy with them, help them when we can, to make their life more pleasant. And for human beings, love also means sharing, warmth, complicity and laughter. True love, they say, is neither possession nor passion.

But, according to the cynics, when it comes to romantic love, birds only experience it in the context of reproduction. Well, so do we. The lover with his bag of warm croissants is not unaware that a good breakfast may lead to other good things, and a little kiss can help to calm the nerves, as preening doves well know. Our reproductive ends may be conscious or suppressed, and a partnership may be about other things beyond creating a child, but reproduction is still a constant background presence in a romantic relationship. Even the most ethereal love has its roots in our animal nature. And so what if it does? Does that matter? The phrase 'animal instincts' tends to have a negative connotation, but these instincts include the tender affection of a pair of doves and the enduring collaboration of greylag geese.

Furthermore, birds are often more gifted than we are at finding love. Seduction, courtship . . . all these things seem simpler for them than for us: they know very quickly which encounters are going to develop in that direction and which aren't. For us humans, on the other hand, our bodies are well hidden under a thick layer of clothing that prevents us from reading the clear signs of masculine or feminine attraction, and so we spend hours, days, months – or even, in the most trying cases, years – trying to gauge whether or not someone is attracted to us. And if they're attracted to us, do they love us? For humans, this simple sentiment has become endlessly complicated, sometimes painful and almost always unsettling.

We struggle to follow our instincts and intuition, or else we overthink our decisions and feelings. Or we're clumsy, not daring enough, or daring too much at the wrong moment, often confused about what we actually want, torturing ourselves as we decide who should make the first move. Then, if we face bitter disappointment, our egos are bruised by these failures and we swear never to love again. Some people are like the wood frog, which can stop its heart when it needs to hibernate through the frozen winter – after a romantic disappointment, they stop loving and never want to become attached to anyone again for fear of being hurt once more. But birds' hearts, by contrast, never stop beating.

We humans lack simplicity and serenity. Blackbirds don't spend three hours deciding whether or not to go and serenade the pretty female: they just do it. Either she likes it or she doesn't, and neither bird seems to think it's the end of the world if it doesn't

8.
LIVING LIFE TO ITS FULLEST

The philosophy of the bathing hen

IF WE WATCH HENS GOING about their daily business, we may be surprised to notice that, in moments of intense satisfaction, they emit a growly sort of purr. A hen is especially likely to make this little growl of pleasure when she's taking a dust bath. It is a sacred moment of the day for her, allowing her to get rid of parasites and look after her plumage. For a bird, having clean, well-groomed, waterproof plumage is a necessity and a matter of survival. Seeing a hen take a dust bath gives us a sense of what one of the greatest pleasures in the world can be.

First, the hen chooses a patch of loose, dusty earth. And then, quite simply, she revels in it. It's hard to recognise the hen in this shapeless mass of feathers – one claw here, one there, a wing here, a head there – throwing up showers of earth in order to cover herself entirely with it. Occasionally, the dust cloud becomes still. The bird sits, eyes half shut, blinking slowly. She utters her little growl of pleasure. This pause can last for an age; the hen has all the time in the world. She seems to bask in the warmth of the sun on her feathers. Then she begins again. She rolls around, plunging her head into the loose earth, making clouds of dust with her wings in endless complicated movements. A second hen comes to observe the spectacle and joins the first. She comes and lies next to the bather for a moment, and they stay like that – huddled together, motionless, pressed against the earth. Then the second hen jumps up and runs off in pursuit of an insect. The first doesn't budge from the hole she has made.

The behaviour of a bathing hen perfectly encapsulates a centuries-old philosophical discourse: *carpe diem*, or 'seize the day'. It proffers the invitation to be mindful, or 'present', the Buddhist prompting to be in the 'here and now' and the psychologist's advice to 'live from day to day', far from memories of the past or worries and hopes for the future. Here and now, the hen takes her bath, warmed by the sun in loose, fresh earth, in the shade of a cherry tree whose green fruits are beginning to form. Here and now, the hen growls with pleasure.

Anthropomorphism?

No. The hen we are observing is no different from us. Neither in the atoms of her body nor in the emotions that she feels. She doesn't see the world as we do, of course: she's a hen. Her senses are not developed in the same way that ours are. Some of them are sharper: she can be on guard, completely silent, aware of the presence of the neighbour's cat long before we have seen it. She gives a cry to warn of its approach while we barely notice the passing feline. We and the hen are two different species. But we are both living things and we share in the enjoyment of feeling the sun on our bodies and of taking a bath – she in dust, we in a warm tub.

The hen's bath should give us pause for thought. Why don't we bathe with the same intensity of purpose? Our lack of plumage means that we don't need to spend so much time cleaning ourselves, but even so . . . Dogged as we are by duties and commitments, worries about the past, the future and the sense of being in a hurry – always in a hurry – we rarely find a moment to experience true delight in the act of cleansing ourselves. The hen does not wash if she is stressed. No, she doesn't take her usual jubilant bath, but either sits still and silent or rushes around screeching. But we still wash even if we're worried or tense, so how can we manage to savour the moment, as the hen does?

The hen teaches us the joy of the present moment. 'Cooot?' She waddles a few steps. 'Cooot?' She jumps. Then, suddenly, she runs off after a white butterfly: too late, it flies too high and fast. But the hen doesn't get upset – she's already moved on to the next thing. Scratch, scratch, scratch she goes, tossing up vegetation behind her. A peck, a look and a peck. What has she found on the ground? Some tiny, unknown race living down there, invisible to our eyes but delightful to hers? The hen is an active being. Conscientious and bustling, she searches, scratches and moves about. But she also knows how to lounge under the trees for hours. She is here, she is present. '*Carpe diem*', she tells us.

9.

HOW TO ADD BEAUTY
TO THE WORLD

The dance of the bird of paradise

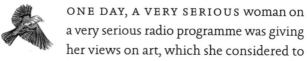 ONE DAY, A VERY SERIOUS woman on a very serious radio programme was giving her views on art, which she considered to be the unique preserve of human beings. According to her, anything beautiful created by an animal cannot be considered 'creative', but rather an interpretation of beauty by human eyes. This beauty cannot have been created by conscious *thought*.

When a long-tailed tit builds its nest, a superb little ball of feathers, catkins, delicate stalks and tiny pieces of lichen, we humans may well be bewitched by its beauty, as if we were observing a work of art. But the two adult birds built their nest to shelter their future offspring, not to showcase for our admiration. If we happen to find it beautiful, so be it. What we don't know is what these little long-tailed tits think about their own nests. How do we know that the females don't build the prettiest nest they can? What is to say that beauty does not form part of their selection process?

Even more pertinent is the case of bowerbirds. This family of Australian birds are beyond masters in the art of nest decoration. Take the satin bowerbird: the male has midnight-blue plumage, and for him, blue is the quintessence of beauty. Is this because of his feathers? Who knows? Like all bowerbirds, he builds an elaborate nest made of small twigs and grasses woven together to form a bower that sits on the ground in a little clearing. To make it still more attractive to his female, our male embellishes it with a sort of blue 'paint'. He makes this from purple, blue or black berries that he mixes with his saliva and charcoal from forest fires. He uses a little piece of bark to spread the mixture on the bower walls. He also places objects around the entrance – blue objects. Bottle tops, berries, pens, flowers, lighters, pieces of plastic – anything, as long as it's blue. Stones, too, are collected, which he carefully arranges according to size, the largest at the front. This creates an optical illusion, causing the stone avenue to appear larger than it is, from the perspective of the watching female. If all this is not art and creation (with the end goal of seduction), then what is it? Why spend hours decorating a nest blue if not to please the eye?

Yes, but none of this is the result of *thought*, the woman on the radio will tell us. Perhaps, but could we not say the same of many things created by humans? Spontaneity and creative drive are sometimes at the root of works of art praised by connoisseurs. Sometimes artists themselves are surprised at what they have created in the throes of inspiration. And even if birds aren't *thinking* in the same way we do, could they not be generating art and creating beauty in their own particular way?

In any case, what is people's motive in creating art? Is it not partly to please others? Musicians, painters and poets – whether men or women – often compose, paint and write inspired by a muse whom they hope to seduce. Is our mode of creation so far from nature and our animal urges? Didn't Freud link art to a sublimation of the libido?

Birds, with their taste for beautiful plumage, charming song and sophisticated nests, come closer to being artists than, say, earthworms do. Beauty seems to be an important, even essential, driving force for many of them, and although it may simply be due to evolutionary selection, their choices sometimes seem to be guided more by what is beautiful than by what is practical. The peacock's tail is splendid but very impractical. The peahen chooses her male according to criteria of pure beauty. The male's strength could be demonstrated by other means than his sumptuous appearance, so why all this abundance of beauty among birds if they are not receptive to it?

One more example: the fabulous birds of paradise. The males perform astonishing dances as part of their courtship parade, with choreography that the principal dancers of the Paris Opera would be hard-pressed to reproduce. And the dance is supplemented by the male's impressive plumage, which boasts patterns and gradations of colour that seem a highly refined form of art.

Obviously, the birds did not 'create' these amazing juxtapositions of tones. Or did they? All of this is, as we have said, the fruit of evolution – but also of natural selection. It isn't hard to imagine that over the course of millennia those individuals best able to perform their dance while balancing on delicate branches over thirty metres above the ground, or those with the longest or most brightly coloured feathers, were the ones who survived. There is no conscious decision behind all of this, but the male's ability to seduce females with his plumage and dancing has a direct impact on reproduction (and thus on the survival of the species).

This 'naïve art' has been exploited by humans: the Papuan people of New Guinea have used the feathers of the bird of paradise in their ceremonies and festivals for millennia. And the men of some tribes decorate themselves with the feathers in order to seduce their partners.

There is an art to plumage. There is an art to birdsong, too. There is no doubt that birds are unrivalled musicians. But are they musicians without knowing it, or are they conscious of their art? We need only go for a walk in the forest in spring to find part of the answer. When two males of the same singing species are close together, we can clearly hear them trying to outdo one another in the power and variety of their songs. The song thrush is a prime example. A lone male sings a beautiful, broadly conventional, song with some minor variation. If another male appears, however, the first one will invent new phrases to enrich his repertoire and will sing more loudly. What is more, in many species the best singers are those who are most successful in their courtship of females. The art of seduction through song . . .

The woman on the radio might argue that these songs, however beautiful they may be, are simply repetitions of tunes learned from their parents, or that they are innate. But that is far from clear. Take the common starling: it certainly isn't the most talented singer. The sound it emits is more like a modest twittering, or even a rather harsh gurgling, than a melody. Nevertheless, the male is quite capable of embellishing his song. Like almost all species of bird, a male starling sings to stake out his territory and seduce, so he incorporates noises that he hears around him in order to please his female and demonstrate his great musicianship. This is why we sometimes hear the sound of a car horn or a mobile phone ringtone from atop a television aerial or the branch of a tree. We look up, puzzled, only to discover that it is a particularly inventive male thrush letting rip!

When we talk of art, do we only mean the ability

to produce it? Or do we also mean the ability to love it? Birds, like many animals (and even plants) can appreciate human music. Our ears are not the only ones to enjoy a pretty tune.

As scientific research on animals has progressed, we have become aware that they are more intelligent, more sensitive and more capable of empathy than we had thought before. So why insist that they have no possible relationship with art or taste for beauty?

Our human art developed from a natural capacity to appreciate all that is pretty and to weave together sounds and rhythms or colours and materials: an ability that birds share. To be artists, we must first master the ability to observe our surroundings accurately, to follow the movement of the leaves or the shifting poetry of the clouds. Then we must transpose all this beauty and the emotions that it inspires into a work of art. Animals may not be able to accomplish this transposition, but who knows whether, on a warm spring day as they gaze at blossom-covered trees, they do not feel the same happiness that we do? How can we be sure that because they are animals they are not sensitive to these things? We ourselves are not all equal in our capacity to perceive and transpose the charm of our surroundings. Those of us who are not very sensitive to art find it hard to understand how others are able to draw or play music or to appreciate the end products. And artists sometimes feel that their sensitivity and mode of self-expression are misunderstood. Are we really qualified to know exactly what is going on in the mind of a bird that has decided to make the loveliest blue nest he possibly can? Or a bird that has given his all to build a nest that he thinks is beautiful, only to have the female turn her beak up at it? Do we know what they feel?

Artists frequently say that they could never be happy if they were prevented from expressing themselves through art. Does this not indicate that there is something innate and fundamentally instinctive in the need for art?

We often hear people declaring, with pained certainty, 'Oh, I'm no artist!' But perhaps we're all creative in our own way. Perhaps we're all sensitive to some form of beauty. We often create limits for ourselves, sometimes because we have been discouraged as children, sometimes because artistic activity isn't deemed worthwhile by those around us. But just because we don't know how to draw doesn't mean we can't make pottery, play music or create wonderful food filled with a poetry of its own. There is an art to suit us all, a creative force as yet unexpressed, just waiting to spring forth. There is no doubt that, like birds, we can all add something beautiful to the world.

10.
HOW TO BE FREE

Opening the cage

SHOULD WE SET ALL CAGED birds free? Is it better to live an unfettered but dangerous life than a safe but limited one? Is it better to pay the price of liberty, or to stay in a gilded prison?

We all know what happens to a canary whose cage door is opened: it flies away, elated, but quickly returns, frightened and eager to regain its familiar world. Anyone who has rescued a battery hen will have seen the same thing. The poor animal, hitherto confined in a cage so small she could hardly turn around, is stunned by her freedom. She starts by exploring only a few square metres around her, close to a wall or hiding behind a pile of hay. She will spend weeks gradually venturing further and further afield before she finally regains the liberty to move around as she wishes.

Still, it would be wrong to claim that some birds prefer to live in cages. When birds have become conditioned to living in captivity, too much freedom all at once frightens them. They fear for their safety and are afraid to explore a new environment. Do we not feel the same sometimes? What if we were to take a child, or even an adult, who had only ever lived in a high-rise block of flats, and released them into the forest? Wouldn't they come running back, begging to be taken home? The same is true in a less literal sense. Moments of great freedom in our lives can be terrifying for some: going on holiday or entering retirement can be difficult. What to do with all that time, and the sudden absence of constraints and structures imposed by others? People don't always want to be absolutely free. True freedom can induce a state of anxiety, in individuals and in whole societies. We want freedom above all else, but we also fear it.

We don't feel the same thirst for the open sky as migratory birds do. They are a potent symbol of liberty with their ability to take off and escape, to travel and reach inaccessible places. Let's not forget that our ability to fly – with the aid of machines – is a very recent development in our history. For a long time we looked up at the birds without being able to join them, thinking ourselves so superior and yet unable to compete with them on their terms. Humankind has always envied the ability to fly and the freedom of movement it brings. We have finally conquered the sky, but we took our time about it!

We humans often find it hard to grasp our own freedom, or to accept that of others. Children today are coddled far more than any chick in its nest. Parent birds let their young do things for themselves and encourage them to fly independently. In the twenty-first century, we rarely see children running and playing on their own in the streets. They are under permanent surveillance because we are so frightened of something bad happening to them. The same goes for couples. Our partner's freedom often scares us, whatever form that freedom may take. Worse still is the family. Here, judgements and set codes of behaviour often rule: everyone must fall in with the group. If one member steps out of line, they are called to order, or are marginalised.

However, the more we fence people in, the more likely they are to escape: the teenager whose life is too restricted will rebel violently against their parents, the woman dominated by a jealous husband will end up fleeing, and the family will be broken.

If we watch birds carefully we see that when hens or doves are allowed complete freedom, they don't stray far from their coop or dovecot. They take refuge there in bad weather or when they sense danger. In this way, they have the security of a roof over their heads, and food and water, but they can spend the day as they please, independent in their movements.

The same goes for humans. If we're free, we don't run away. If we're happy in the fold, we will always return to it. The best way of keeping or protecting someone is surely to make their nest welcoming so they will want to come back to it. Perhaps that's what we need: a harmonious balance, blending the advantages of domesticity and liberty.

11.
A QUESTION OF FIDELITY

The curious life of the dunnock

THE GARDEN HEDGE IS A refuge for birds. Large numbers of them shelter inside it in winter, flitting back and forth to nearby feeders, fighting noisily over provisions. But down on the ground other little brown birds are pottering about unnoticed. These are dunnocks, also known as hedge sparrows for their habit of sticking close to the bottom of the hedge. They might just as well be called invisible sparrows, given their uniformly drab plumage: brown on top and blueish grey underneath. They blend into the background wonderfully well and only their call, a piercing, tremulous '*tsii*', gives them away. Even then, only a practised ear can locate them. In short, the hedge accentor, to call this bird by another of its names, is the very model of an unremarkable bird. Nothing in its plumage or song excites the slightest human curiosity.

We shouldn't judge by appearances, though, for this apparently monk-like creature is concealing something beneath its habit: a dissolute life. In reality – and contrary to the long-held belief that birds are faithful partners – the dunnock is an expert in polygamy and polyandry. In theory, one male and one female build a nest together and raise their young. Until recent studies proved us wrong, we believed dunnocks to be models of conjugal fidelity. But it's all just a front! Things are actually rather different. Both male and female dunnocks are consummate swingers. Each male has his own 'missus', but will also mate with any willing female who passes through his territory. The same goes for the females. A lady dunnock doesn't spurn the advances of the neighbour from the other side of the hedge, and sometimes she even makes the first move. Let's face it, there's a bed-hopping farce being played out at the bottom of your garden!

Humans, it turns out, have invented nothing new. We need only leaf through the book of biodiversity to see that there is an array of sexual licence at all levels in the animal world that would bring a blush to the cheeks of the most uninhibited *Homo sapiens*. We surround ourselves with an aura of virtue, unlike animals, which make no excuses for their behaviour. Except, that is, for our dunnock, which conceals its dissolute life beneath a cloak of banality. It hides its true colours well: no seductive plumage, no showy song . . . like so many people who seem very conventional and yet turn out to be happy libertines.

But the lurid details of the dunnock's sex life don't end there. Studies have also shown that in times of plenty the males lead the charge, while in times of famine, the females do. This is because when food is plentiful, the females tend to remain within a small area (because there's no need to go far in order to find it) and it then becomes only too easy for the males to go about their promiscuous business, thus increasing their chances of reproduction. When food is scarce, however, the birds must increase their territories if they are to find enough food and the females cover larger distances, in turn increasing *their own* chances of meeting a male. In general the female birds limit themselves to two mates, the stronger of which is the alpha male. But, as everything in nature is nicely balanced, when food is relatively easy to find and territories are at their normal size, the dunnock chooses to lead a 'conventional', monogamous life. So is infidelity a sign that something, somewhere, is a little off-balance?

Dunnocks do seem to expend a lot of energy 'making the beast with two beaks'. Everything they do serves a specific purpose, though. Before mating with a passing female, the male pecks her cloaca, the sex organ in all passerine birds (commonly known as songbirds). It is a strange form of foreplay, which surely can't be pleasant for the female, but by pecking hard at his partner's behind, the male causes her to have a muscular contraction that ejects the sperm of her previous partner! He then mates with her, depositing his own sperm in place of his rival's. What better way to ensure one's paternity? This strange practice is especially common among alpha males, who will go on to produce the heartiest chicks and will also take the most care of them, thus increasing their chances of survival. The selection process is highly effective.

Does the story end there?

By no means! The female has the last word. Unbeknownst to her partner, who thinks he is the sole father, she is actually able to retain part of the sperm from prior exchanges, preserve its genetic material and produce broods containing the offspring of several fathers. A superb example of natural selection.

This practice, known as polygynandry, thus serves a definite purpose among dunnocks: it ensures large broods and wide genetic diversity. Unsurprisingly, the species is widespread, present throughout Europe, Scandinavia and North Africa.

What about us? Are we polygynandrous? Promiscuous people may practise a form of polygynandry, but the similarity ends there. Or does it? Scientists have proposed the idea that the shape of the human penis has a specific function, arguing that the movement of the tip during repeated back and forth movements in the vagina removes the sperm of any rivals. Whether or not this is true, men have always expended a great deal of energy finding ways to perform the psychological equivalent of cloacal pecking – in other words, stopping their female partners from being impregnated by others. These range from exaggerated censure of female sexuality to social pressure and telling stories to little girls about a single prince charming who will save them. Not to mention the horrifying chastity belt, or genital mutilation. These things aren't a million miles from a cloacal peck . . .

By way of conclusion, let's finally put to rest the notion that the majority of birds are faithful for life, paragons of a romantic ideal of monogamy. Of course, there are monogamous species, such as geese, swans and certain birds of prey, but for other birds the reality is more subtle, and when we study them we encounter all possible variations of relationship, ranging from monogamy to polygamy and everything in between. Their behaviour often depends on the circumstances: the environment, the availability of partners and food, and other factors. In a word, birds *adapt*. We have no idea if birds are furious or sad when they discover an infidelity. Do the feathers fly?

Among humans, including those who see themselves as following very liberal codes of sexual behaviour, the dream of true love often complicates things. Freedom doesn't necessarily mean having multiple partners: we may also feel liberated by being faithful. We are caught between ideals and reality and, in the end, sexual love, whichever form of it we pursue, is often a

12.
DID CURIOSITY KILL
THE . . . BIRD?

The audacity of the robin

FROM BIRTH TO DEATH, a bird's life is one long series of risks. Flying, finding food, reproducing, raising young: everything is dangerous. Without risk, though, there is no life, and while some birds' natural curiosity may lead them to take risks, also enables them to discover new possibilities: rich food sources, perfect nesting locations or ideal places to rest in safety. He who dares, wins, and for animals, especially birds, curiosity is an effective and often sophisticated way of adapting to the environment. It is also, therefore, an essential tool for survival.

Take the robin, a familiar bird to all gardeners in Western Europe and one that, as we have seen, is especially fearless. The robin will perch calmly on a chair, spade or any other man-made object, within just a few centimetres of a person, to observe them at work. Head tilted and beady eyed, it sits completely still and seems to watch with interest and curiosity. As soon as the rake turns up a worm, the robin pounces and gobbles it up. Then it waits for the next one, its red breast puffed up, seemingly unafraid of the gigantic human being at work beside it.

Originally, the robin was a woodland species that kept its distance from humans. It has always been known for its habit of following mammals such as deer or boar while they graze or dig – sometimes getting quite literally under their feet – always with its head tilted to one side and with gleaming, watchful eyes. It hops along, following the animal; just like in the garden, it's on the lookout for small insects unearthed by a hoof or snout.

The robin continues to be a woodland bird, but is often found at the woodland's edge, close to human habitation. Over time it has gradually come closer and closer to people, to the point of following them around the garden. The robin's curious, opportunistic nature is especially noticeable in Great Britain, where it has become part of popular mythology and is a familiar image on Christmas cards. The widespread British habit of feeding birds in winter has probably contributed to reducing the robin's reticence, but it has also made it one of the main victims of cats. This fearlessness is much less pronounced in countries where the robin is hunted (such as in Southern Europe) or where it remains a largely woodland bird (in Eastern Europe). In these places it remains timid and wary.

The robin's curiosity has certainly allowed it to broaden its diet and extend its territory, most notably into urban gardens where the winter is less harsh than it is in the countryside. An urban robin thus has a greater chance of surviving a cold winter than a woodland robin does.

A naturally curious nature has brought other species

into closer contact with people. We've all seen pictures of sparrows in Paris, blue tits in London and even jays in American national parks, coming to eat out of people's hands. These wild birds have lost all fear of humans. They watch us as we sit eating on a park bench or in a picnic area and come to pick up scraps left on the ground when we leave. Gradually, they become bolder and appear while we're still there. Eventually, one will grab a stray crumb, then perch on the table and, finally, eat out of our hand.

Taking risks can also be a survival strategy. Many birds take risks when faced with a potential predator in order to scare it off. An individual might approach the predator, emitting alarm calls that bring other birds to the scene, making alarm calls of their own. The attacker is driven away by sheer noise and strength of numbers. When out walking, we are sometimes surprised to see a bird approaching us, seemingly wanting to know who we are, only for it to disappear again into the depths of the forest or the marshes. Their curiosity could simply be a way of gauging whether we are a potential threat. If we humans were kind to all animals we would surely have a more intimate relationship with them, and they would see us as a fellow species, not a threat. On the uninhabited Madeiran Desertas Islands, a little bird called Berthelot's pipit has no fear of humans whatsoever; it will immediately come and perch on a shoe or knee to collect a crumb, glancing about curiously all the while.

Curiosity does not, then, kill the bird. It is a natural behaviour and a powerful tool in the bird's evolution-ary arsenal, allowing each species to survive and thrive. This quality is also innate in humans. We too make use of curiosity as a creative force; it pushes us to explore new continents, to go to the moon or discover cures for diseases. It is at the heart of everything that drives us to evolve. It's a good thing to remember the next time we notice a robin observing us from a nearby fence post.

13.
WHY DO WE TRAVEL?

The Arctic tern and the call of the sea

MIGRATORY BIRDS EMBODY the spirit of travel. What makes them take off one fine morning towards new horizons, only to return a few months later?

> *See them as they pass! The truly wild ones.*
> *Following their desire, far from earthly bonds,*
> *Over woods and seas, and winds, and mountain crests.*
> *The very air they drink would shatter our poor chests.*

This is how the French poet Jean Richepin described migratory birds, his lines later set to music by the popular French singer Georges Brassens. Many of us probably share Richepin's fellow feeling for these free, single-minded voyagers who cover such unimaginable distances. The Arctic tern is one such bird. As its name suggests, its breeding grounds are in the High Arctic, from Siberia through Northern Europe and into North America. As summer ends, it leaves for the Antarctic and sub-Antarctic seas, the site of its winter wanderings. Overall it travels up to 55,000 miles in one year. A bird that lived for twenty years could thus travel more than four times the distance from the earth to the moon in its lifetime.

The Arctic tern is an eternal voyager, a follower of light. It is said to see more daylight and sunshine than any other species: the endless day of the Arctic summer followed by the endless day of the Antarctic summer on the other side of the world. The blue oceans and distant white shores are the only world it knows. In the northern summer it lands on the tundra, a small oasis of vegetation and bright flowers where it spends a few short weeks.

But why does this bird travel? Is this one long holiday? It could end its southward journey when it reached the temperate European coast, or only push as far as West Africa. But no, it crosses the equator and continues through the strong westerly winds of the southern hemisphere, known as the roaring forties, and then wanders around the Antarctic before moving north again. The exact reasons for this journey – the longest made by any living creature – are not fully understood, but recent studies have suggested that the tern may choose these regions for their abundance of extremely nutritious plankton. The rest is probably the result of several millennia of evolution.

In the end, perhaps it's not so hard to understand why the Arctic tern travels so far each winter. After all, we humans undertake our own seasonal migrations – known as holidays – at least twice a year if we're lucky. What drives us to leave our usual habitat? The need for a change of scene, for new horizons; a little taste of the unknown; the desire to lose ourselves elsewhere in a culture unlike our own; to forget the day-to-day and break with the monotony of our usual rhythm. All these factors inspire us to travel.

People, like birds, are either inveterate travellers or confirmed homebodies; those who answer the call of the sea or those who stay put; those who have the 'thirst for the skies' described by Jean Richepin or those who like the comforts of home. The tawny owl has no wish to leave the forest where it was born, but swifts and swallows have only one desire as soon as they are fledged: to take flight.

The child who stares wide-eyed at his surroundings as he travels with his parents, drinking everything in, is very likely to grow up to be a footloose adult. The more he discovers, the more he will want to discover – all those thousand obscure corners of the world, all those far-off places that he could only dream of when he used to gaze at the atlas, never imagining that one day he would set foot in the Syrian Desert, the Caucasus Mountains, the Korean islands or the Brazilian jungle . . .

We, like birds, are shaped by the voyages we undertake when we are young. The young Arctic tern learns its route when it migrates for the first time, and then follows the same path as an adult. For human children, looking at an atlas, watching documentaries about distant places or going on holidays with parents are all rites of passage into a life of travel.

Every journey changes us a little. It makes us see the world through fresh eyes and provides a remedy for isolation, withdrawal and fear or disdain of others. All journeys teach us solidarity, like migratory birds that support one another during their long flight by constantly calling to those around them. We never return from a journey the same: we leave a little of ourselves behind and bring a lot back with us. New horizons make us grow, give us more heft and substance. Encounters with others open us up to everything around us: different people, environments and ways of life. Above all, travel teaches us about ourselves, about what we are capable of in terms of endurance, discomfort and how we adapt to challenging situations. On the other side of the world, in a different time zone, fatigue kicks in and the mask comes off; it's the best way to reveal our true selves. And perhaps that is what we seek when we travel: to learn our own truth.

14.
POWER GAMES

The crow and the vulture

A DEAD COW LIES IN a pasture high in the Pyrenees. Two or three large crows circle above it for a long time and then land on the carcass. They begin to attack the animal with their beaks, but soon a griffon vulture, with its intimidating wingspan, appears. It approaches the dead animal with a rolling gait, spreading its wings to make it seem even larger than it actually is; the crows immediately make themselves scarce. Then, one by one, other vultures arrive. By dint of outstretched wings, pecks and guttural croaks, a precise hierarchy is soon established in which each bird has a clear position. The vulture that one moment ago held the prized location near the victim's entrails now finds itself ejected from the front row of diners and must wait its turn to come to the table. It has only the crows, now completely sidelined, as companions in its misfortune. Everyone must know their place at this impromptu banquet. But, should a sharp-toothed fox appear, the assembled company will give way and let it gorge itself at leisure. When the dominant animals are sated, the less dominant will have their turn, followed by the not-at-all-dominant. By this time, even the crows will feel confident enough to join in. While the first to the feast digest their over-hasty and over-indulgent meal, the followers enjoy a calmer feast of leftovers exposed by the dominant animals, who did all the hard work. Now that is something worth waiting for.

In the completely different environment of the marshes, ruffs behave in a similar way. As their Latin name, *Calidris pugnax* – warlike shorebird – suggests, these waders (who live near mudflats) are quarrelsome little birds. At least, the alpha males are. As soon as spring begins, these males begin to sport a flamboyant ruff of red, black, grey and white feathers. They spend most of their time fighting over females in a special 'arena' set apart from the other birds. The females stay on the sidelines looking for food, largely indifferent to the spectacle. A group of beta males lurks close by; they have not achieved dominant alpha status and only have a modest ruff that is usually plain white. Are they observing passively as the alpha males puff up their feathers and attack each other with gusto? Yes and no . . . While the dominant males are caught up in their posturing, some of the beta males are quick to seize the moment and enjoy a surreptitious tryst with one of the nearby females!

We all know someone a little bit like these swaggering alpha males, someone who uses their biceps – actual or figurative – to impress the crowd. These show-offs, whether political, intellectual, professional or sporting, only have power over us if we accept that power. But if you keep your distance, follow your own path and observe the person who expends all their energies imposing their will on others, you'll see that they miss out on life's pleasures. Your life will be less competitive than theirs, and all the more interesting for it. Think of the one hen in the coop who uses her beak to show the others who's boss instead of focusing on the more important business of eating. This is what dominant hens often do, becoming so absorbed in power games that food is swiped from under their noses by the riff-raff, who are more focused on their stomachs.

What exactly do we dominate when we achieve power? More often than not, real power is held by the people pulling the strings in the background. This discreet person may act quietly but achieve more. Hierarchy is often a game – we clamber to the top of the pyramid only to be quickly deposed. The world of birds and mammals is full of examples of a dominant male battling endlessly to reach the upper ranks of the hierarchy only to find himself so exhausted by his efforts that another takes his place almost immediately.

The thirst for the recognition and power that we so cherish can lead us down paths where we may easily get lost. The closer we are to the summit the less we notice the details, as the little things that are the spice of life fade from view. Are political leaders, celebrities and bankers really so happy? They may have reached the peak of career glory, but won't there always be some other person just waiting to knock them off their perch, or a beta male or female who will make their lives miserable? An alpha-male ruff might stand a good chance of attracting a female, but does the same go for humans? Socially independent woman are now in control of their own lives and loves. Will they choose a boastful alpha male who is likely to run off and have an affair? Scientific studies have shown that beta-male birds are more dependable and attentive to their mates. Why? Perhaps because they don't have the luxury of choice.

Some birds, such as another wader, the dunlin, have clearly understood this. Female dunlins tend to favour smaller over larger males. Why? Because the smaller birds are better at defending their territory and young than larger birds that are less agile in flight or nimble when faced with predators. Perhaps there are lessons for us humans here as well.

So, is it better to choose a beta or alpha male? Each to her own!

15.
SIMPLE PLEASURES

Happy as a lark

THE WIND WHIPS ALONG the Brittany coastline at Morbihan. Hats fly and hair dances, cheeks glow in the cold. The gulls in the port seem to be having a great time, looping, gliding and landing briefly only to take off again. These aerial acrobatics don't appear to serve any particular purpose, but they do look like fun. Perhaps this is what happiness means for a bird: playing in the wind.

Dusk in the Cantal, in central France. As the last rays of the setting sun slant through the warm, calm September air, water spurts up from the path like an out-of-control sprinkler. Thousands of droplets fly, glittering in the sunlight. They're coming from a puddle where dozens of starlings are having a bath. They are all piled in together, beating their wings, splashing water into the air, looking as though they're trying to empty this poor little puddle. Before they descended, it was quietly reflecting the sky, now it's a starling paddling pool. Seeing them splashing madly like children at the swimming pool, it's hard not to wonder if this is what happiness means for a starling: an invigorating splash with friends.

Swooping seagulls, splashing starlings, doves warming their wings in the sun, a pleased-looking blackbird parading along a wall having swallowed a big worm, a sleepy heron with its eyes half-closed and feathers puffed up, standing on one leg . . . birds give us many glimpses of well-being, playfulness, tranquillity and frivolity. What is the definition of happiness for them? Having a full stomach? Being safe from predators or going about their daily business without fear of danger? For them, happiness starts with the absence of unhappiness – just as it does for us.

Can birds be depressed? Can they feel pessimistic or bitter? Not when they're in their natural environment, or at least not for long: sadness and discontent are often linked to brooding over the past or worrying about the future, but birds live in the present. That doesn't mean they don't sometimes show signs of distress, such as when a monogamous bird loses a partner or when a parent bird loses its nest, eggs or brood. We don't, however, know what they're feeling or how long it lasts.

A caged bird kept in poor conditions may show signs of 'depression' – dull, deteriorating plumage or general dejection. Some species, such as the South American quetzal or the osprey, are especially ill-suited to captivity and its effects may even prove fatal. Other species react differently, however, and they may reproduce in captivity (stressed birds never reproduce) and even live longer than they would in the wild (whereas a stressed animal's life expectancy is significantly decreased). It would seem, then, that having a decent, albeit restricted, living space with good food, fresh water, regular care and protection from the 'stress' of predators can sometimes compensate for the loss of freedom.

Establishing what makes a bird happy is far from straightforward. And how can we define happiness? Many philosophers have tried, often suggesting that happiness can be attained through an ideal combination of wisdom and self-control. Are birds Epicureans, understanding the need for moderation and contenting themselves with small pleasures without succumbing to the kind of excess that can lead to suffering? This is often the case. A wild bird eats as much as it needs, guided by a certain natural abstemiousness . . . most of the time.

There are always exceptions to the rule: thrushes, for example, are not models of temperance. They are well known for feasting on fermenting autumn berries and getting completely intoxicated. The alcohol in the mature fruits, consumed to excess, makes them extremely drunk. They can't even fly in a straight line. It's quite a sight! No more Epicurean moderation, this is more like what Rabelais would have recommended as a recipe for happiness: enjoy the moment and get stuck in.

On reflection, birds can be rather hedonistic: individual pleasure is their ultimate aim. They are Rabelaisian *bon vivants*, avoiding displeasure and, when the occasion allows, stuffing their faces. If there is one philosopy that birds really have no concern for, it's Stoicism. Living in the moment as they do, they have little need to achieve self-control and renounce desire.

The truth is that birds don't question happiness. They live it. When things are going well, they're happy. It's as simple as that. Knowing how not to worry – perhaps that is the beginning of happiness.

16.
WHAT IS INTELLIGENCE?

Bird brains

WE HUMANS LIKE TO MARVEL at our own intelligence, believing that our powerful brains make us superior to all other living beings on earth. We also love to rank animals according to their intelligence: those judged 'clever' (dogs, dolphins, great apes, etc.) are worthy of respect, those judged 'stupid' are not. Birds, much like most fish, are often seen as being a bit dim. But they aren't. And anyway, does it really matter which species or which individual is more 'gifted' than another?

The business of defining intelligence is a complicated one. There are IQ tests, of course, but they ignore a lot of what makes up a human being. How highly would Vincent Van Gogh have scored? Or Apollinaire? Artists employ forms of intelligence that can't be measured by logic tests. Intelligence is the ability to understand, it's true, but to understand what? The way a motor works or the beauty of the world? Fluid mechanics or the emotions of the person in front of us? How is solving a mathematical problem superior to composing a poem, and in what way is being a chess champion a better mark of intelligence than the ability to create harmonious compositions of colour or expertly play the violin? The very definition of intelligence is often biased by those who created the criteria. In actual fact, different sorts of intelligence coexist. For example, we are now increasingly aware of emotional intelligence, the kind that helps us to live harmoniously with those around us, with a subtle understanding of human interaction and relationships. Emotional intelligence also draws on the ability to adapt, allowing us to get by in new situations.

How does this manifest itself in birds? We should begin with a caveat: it's important not to confuse intelligence and evolutionary processes. For example, urban great tits sing more loudly than rural ones because the city noise threatens to drown out their song. They must raise the volume if they are to communicate with partners. It is not that the great tit has understood what it needs to do in order to make itself heard, but rather that it has adapted naturally to a changing environment.

But back to being stupid, or not. In America, hummingbirds quickly learn that it's easier to get sugary water from a bird feeder than from flowers, and they become so addicted to it that people are advised not to put sugar water in their feeders, or at least only occasionally. Meanwhile, pigeons, those much-derided residents of our cities, have been found capable of understanding the concepts of time and space.

Corvids are ranked unusually highly for birds on the animal 'intelligence' scale. This family includes crows, ravens, rooks, jackdaws, magpies and jays. Take the jay, known for its ability to plan ahead. When autumn comes it collects a large stock of nuts and seeds and hides them here and there, ready to be eaten during the winter when food is scarce. (Incidentally, it hides so many that it sometimes can't remember where they are; these forgotten caches are a welcome treat for other animals that come across them, but they also benefit the forest itself, as the jay's buried stores help to disperse seeds and plant new trees.) And the jay has another trick up its sleeve. If it is hiding its seeds and notices that it is being watched by another jay, which might well come and steal its precious trove, it starts to change its behaviour. It pretends to hide the grains without actually doing so, in order to trick the potential thief. Crows are capable of similar pretence.

There have been many studies of corvids, each with fascinating results. It has been discovered that in the wild, some crows can use a tool to get to food that is out of reach, utilising a twig or small branch, just as chimpanzees do. Some corvids, such as New Caledonian crows, are even more advanced and can use branches as hooks. In a lab environment, a crow has also bent a small wire to make its own hook just right for grasping food.

Some urban crows have even managed to turn city

life to their advantage. They have learned to make use of vehicles and traffic signals to access their food: they take a nut to the spot where cars stop at red lights for pedestrians to cross. When the lights turn green, the crow drops the nut and it is quickly crushed by the cars. When the lights turn red, the crow flies down and begins to feast. Until the next green light . . .

Other studies have suggested that crows can also teach each other how to make tools or how to use particular strategies. They are thus capable of a certain degree of knowledge transmission, an ability also found among great apes but that was, until recently, thought to be unique to humans.

Finally, there is the experiment with the magpie which, when placed in front of a mirror, recognises its own reflection. When researchers put a red mark on the bird's forehead, it sees it and tries to scratch it off. In this way, some birds (crows and parrots, for example) can pass the self-recognition or mirror test. A human baby can only do this from the age of eighteen months.

So birds aren't stupid, and intelligence and self-recognition aren't unique to humans. It was thought for a long time that the larger the brain, the more evolved the species (and, by the way, men used this theory to develop the obtuse argument that they were superior to women, whose heads and brains are generally smaller). Birds give us a clear counter-example: compared to a monkey or an elephant, a crow does indeed have a tiny brain, but one that contains twice as many synaptic connections as that of any mammal. Here is our proof that brain size is not important.

We humans are quick to conclude that anything that is unlike us is inferior to us. (Indeed, some people spend their time seeking proof that the Other among our own species is inferior, whether because of the colour of their skin or because of a physical disability.) Animals have thus always been considered inferior. The English word 'beast' has the same origins as the French word '*bête*', which means 'stupid' as well as 'animal'. However, in our enthusiasm to measure and classify everything we use distinctly 'human' criteria to define animal intelligence, the real nature of which escapes us.

Clearly, humans possess unrivalled intelligence, but all species possess the intellectual faculties most appropriate to them. Birds are better than we are at travelling without getting lost, at locating a predator or finding food deep in the forest. The fact that we'll never see two hummingbirds playing poker is neither here nor there.

Perhaps, then, we should direct a rather more humble and curious gaze towards crows, jays and their relatives and recognise that we still have much to learn about animals. They might be able to help us understand the origins of language, abstract thought, desire, fear, intention and even imagination. After all, the fact that a crow can use a tool or plan a strategy places it in a special category in the animal kingdom; it brings it closer to us. Didn't Aesop, almost 2,600 years ago, write the fable of the crow who throws pebbles into a jug to raise the water level inside so that he can quench his thirst? A mere crow . . .

This human arrogance in the face of animal intelligence and understanding brings to mind Claude-Lévi Strauss's description (in *Tristes Tropiques*) of how people judge one another:

> The white men proclaimed that the Indians were beasts, while the Indians merely suspected that the white men were gods. Given their equal states of ignorance, the second assumption seems more worthy of humankind.

Perhaps the truest sign of intelligence is humility.

17.

BEYOND GOOD AND EVIL

The morality of the cuckoo

HOW ABOUT SHAKING THINGS up a bit? How about banishing once and for all the naïve, cloying image of little birdies that love each other dearly and spend their lives singing sweetly and preening their pretty feathers? This is a persistent fantasy, a chocolate-box vision of sweetness and light.

The reality of nature is harsher and more problematic. The complexity of what we might term morality is linked to the evolution of animal species. For example, snails are not, in their day-to-day lives, subjected to difficult conditions or the need to compete that might push them to extreme behaviour. As long as the lettuce is fresh and they steer clear of the gardener's foot, snails lead a peaceful existence. When it comes to reproducing, they are hermaphrodites, meaning that there are more potential mates and therefore less competition. At the other end of the chain, however, mammals often lead a more aggressive existence, in part due to the increased precariousness of their existence. For some carnivores and monkeys, kidnapping, rape and infanticide are part of life. For humans, too, unfortunately.

What about birds? Which aspects of their behaviour appear to us as 'good' or 'bad'? And how might this human judgement influence our understanding of their lives? Birds are not snails, but neither are they chimpanzees. Still, we find some of their behaviour astonishing. The cuckoo, for example, behaves in a way that can seem scandalous by human standards.

The female cuckoo lays each of her eggs in a different nest that is already being tended by another bird (parasitism), and as soon as a young cuckoo hatches, it pushes out the eggs (or even the chicks) already in the nest so that it can receive abundant food from its new host parent. If we were to describe the cuckoo's child-rearing technique in our own terms, we would surely call it immoral. The parent cuckoo spends no time whatsoever looking after its young – surely a matter for social services!

In fact, the cuckoo's strategy is the fruit of a long evolutionary process that is neither senseless nor intended to wipe out the other species. In the animal kingdom the ultimate aim is to produce as many descendants as possible, and to expend as little energy as possible in doing so. By laying each of her eggs in a different nest, the female cuckoo increases the number of healthy offspring she is likely to produce. This is partly because the parental instinct of the host species kicks in no matter how large the chick they find in their nest, so they will feed it until it can live independently. Furthermore, if one host nest is damaged or raided by a predator, the cuckoo's other eggs are still safe. The cuckoo quite literally doesn't put all her eggs in one basket. The other benefit for the cuckoo is that it invests almost no energy bringing up its young, unlike most other bird species for whom rearing chicks is extremely physiologically demanding.

Let's take another example. A child sits eating a bag of sweets in a town square. Suddenly, two other children come hurtling along, grab the sweets, and run off. The child's mother, who witnesses the scene, gives an angry shout and the child cries. It's an everyday drama with nothing out of the ordinary, and it's not hard to determine whether or not the act committed is reprehensible.

We can observe similar behaviour among birds. Have you ever heard of skuas? Possibly not. They are large, dark sea birds, elegant and stately in flight, that spend their lives acting as parasites and stealing food from other bird species. Instead of conscientiously doing their own fishing, as terns (their preferred victims) do, skuas wait for another bird to catch a fish and then swoop down after it. Following a lengthy pursuit over the waves, the skua forces the bird to surrender its catch. It's an impressive aerial spectacle known as kleptoparasitism, but it would certainly be deemed *morally* wrong according to our system of values. The reasons for this behaviour aren't yet fully understood because, unlike the female cuckoo, the skua expends an awful lot of energy committing its crime. The benefits of the reward must outweigh the cost of the pursuit.

Theft, which is condemned in human society (but nonetheless rife), is a strategy used by many animal species. And what about the victims? Although it frequently has its dinner stolen by the skua, the tern is probably grateful for its tormentor's presence at certain times, especially during the reproductive period. Skuas nest in the High Arctic and guard the sites vigilantly. If a marauding Arctic fox approaches skuas will be the first to raise the alarm and attack the mammal, often causing it to retreat. Terns know that it is in their interests to nest close to the skuas and benefit from their protection during the vulnerable nesting season. When later on the young have flown the nest and it's time to make the annual journey south, a few stolen fish won't be a disaster: it must be a price worth paying.

The distinction between good and evil is deeply rooted in our subconscious and may seem obvious, but morality evolves with time and social change: today's good is not necessarily the same as yesterday's, and although there are some basic universal taboos, evil isn't the same everywhere. What distinguishes us from other species is our nuanced understanding of the morality of a particular behaviour according to rules that we ourselves create and that may sometimes change.

The laws of nature, however, function outside any judgement of what is good or evil. Thus, birds can help us to question some of our assumptions about what we really consider to be right or wrong. During the Nazi occupation of France, those who concealed Jews in their houses broke the law and were told that what they were doing was wrong – but it was right. It's useful to remember that good and evil are neither natural nor immutable; they are a human construct, collective as well as individual, a construct that can change.

18.
SHOULD WE BE FRIGHTENED BY OUR OWN SHADOW?

The flight of the chaffinch

A CHAFFINCH HOPS about on the lawn. Suddenly a shadow frightens it. Is it a cat? The bird panics and flies off. In its haste it doesn't see the windowpane. It crashes into it, breaks its neck and dies instantly; a little ball of feathers in the grass, all because of a shadow that probably wasn't anything at all. Fear may lend us wings, but we shouldn't always use them.

Our own reactions to fear are not so very different. When night falls and we're alone in the house, we too can get the jitters. A bump, a banging door, a strange gurgling from the boiler, the wind moaning in the trees ... We may be grown up, but we still jump and conjure up nightmare scenarios filled with robbers and axe murderers. The child who thought there was a monster hiding under the bed is never far away. How many nights over the years are disturbed by imaginary fears that we are ashamed of and conceal deep within ourselves?

Fear is one of our most ancient emotions. It provokes the same symptoms in people as it does in birds: our heartbeat accelerates, our muscles tense, we become jumpy and cry out, or sometimes we are paralysed. A frightened bird tends to empty its bowels and in moments of intense fear humans may do the same. Fear can be communicated between birds, as it can between humans, and whole groups can be seized with panic. This mass fear can end in disaster, even when there was no actual danger. Just as when the chaffinch crashes into the window, it can be fear itself that causes the catastrophe.

Do birds that sleep at night share our fear of the dark? Nobody knows, but when night falls they quickly find a well-hidden spot in which to sleep. We humans are the only species who have decided to defy nature's rhythms and stay active even when it is dark outside, inventing artificial light in order to do so. Our eyes are not designed to see clearly in the dark, and if we find ourselves alone at night in the middle of nowhere we are far more scared than birds are!

Birds also share some of our other phobias: they are rather claustrophobic, disliking small spaces and some particularly sociable species can't stand being alone and are only really content in a group. In fear, just as in many other things, they are like us.

So what is fear for? Almost all living things are 'programmed' to feel it, and it's useful: it works as an emotional safety barrier, allowing us to prepare for danger. Fear saves us, and we – like chaffinches – should be thankful for a good dose of cowardice.

The most fundamental human fear is the fear of death, and many of our other anxieties spring from this source. It's the same for birds, who are afraid of predators: they are continually on their guard because they know they're on borrowed time. For the most part we no longer need to fear 'predators' (except for other human beings, of course); the chances of finding ourselves face to face with a tiger or polar bear are slim, but we still experience fear, nonetheless.

While many of these terrors spring from real threats (accidents, our children being in danger, etc.), most of our misgivings are the work of our imagination, just as they are for a bird alarmed by a rustle among the leaves. Our brain invents catastrophic, far-fetched scenarios so that we are exaggeratedly frightened of certain things. We won't actually die if we fail an exam, blush in front of the object of our affection, or spend a night in a completely dark room. So why do we react to these things so strongly?

Why, with our sophisticated human brains, are we sometimes as terrified as a chaffinch? Such fear is especially puzzling given that it can ruin our lives, plunging us into a state of stress that damages our health. Fear can cause us to lose sleep, it can diminish our appetite and weaken our immune system. In cases of extreme fear, those with weak hearts – bird or human – can suffer cardiac arrest.

Is there any way of differentiating between good and bad fear, between the legitimate and the irrational?

Birds certainly can't, and we do have an advantage

over them in this sense: we can reflect and put things into perspective. A bird listens to its fear above everything else. It must survive and that means flying away quickly. Humans, however, are most of the time capable of reining in our irrational fears. On the other hand, we have lost some of our animal instinct to sense emotions that are deep within us and to understand that sometimes we should listen to what fear is telling us because its message can be vitally useful: it can protect us.

'Bad fear' stops us in our tracks: it paralyses us. It feeds us excuses not to take risks and stops us from living life to the full. Sometimes, though, we refuse to listen to our 'good fears', our gut feelings – when, for example, we know that a new, well-paid job clashes with our values but we take it anyway; or when we sense that being around a particular person isn't good for us; or that a seller is dishonest, and we stifle that little voice and carry on regardless.

Why is it so hard to trust our reasonable fears and ignore our 'bad' ones? Perhaps we need to listen to our bodies and sensations more carefully, and remember our animal instincts. Our brains can play tricks on us, allowing reason to drown out the heart. The more we are in touch with our emotions, the less we stifle them, then the more we are able to hear what they have to tell us.

19

ACCENTS AND OTHERNESS

Calais chaffinch or Marseilles chaffinch?

BIRDSONG IS A SOURCE of wonder for those of us who are sensitive to music. Each species has its own language: sounds that make it unique.

Recent studies have explored the 'regional accents' that can exist among groups of birds within the same species. Several birds have been found to develop such accents, such as sparrows, crossbills and the common chaffinches found in forests, parks and urban gardens. In the case of the chaffinch, you need only have a good ear and take the time to listen carefully and you will soon notice that its song is different depending on whether it lives in Strasbourg, Paris, Ajaccio or Pau. Yes, a Marseilles chaffinch sounds different from a Calais chaffinch! The basic phrasing is the same, but there are different flourishes and extra notes added on at the end. This is why such variances have been described as accents or dialects rather than entirely different songs. It isn't clear why such distinctions exist, but it may be that by developing specific regional songs, chaffinches are able to recognise individuals from certain places and perhaps avoid chaffinches with 'foreign' accents joining the group. We can't be certain of the reasons, but we do know that a chaffinch that moves to a different region modifies its accent in order to blend in, just as we do!

In the street, in a café or on a train, our ear will immediately pick up on an accent that isn't from 'round here'. We find some accents pleasant or even seductive, others unattractive, while some make us laugh. Is it simply a question of taste? Surely not. Our collective imagination influences how we perceive an accent. For example, French ears tend to enjoy the sound of the Quebecois accent, as heard in far-off Canada, whereas many of us tend to dislike accents that we hear closer to home. It's hard to escape judging or being judged by this kind of criteria: the area in which we are born and rasied has a strong influence on us. Some of us feel happy with our own particular accent. Others, on leaving home, may try to lose their regional accent and, like the chaffinch, adopt the accent of the place in which they are currently living.

Just as chaffinches will eject an intruder with a slightly different accent from their own, we may (albeit unconsciously) have our own internal in warning light that flashes when we hear an alien accent. An accent seems to communicate more to us about a person than they themselves are willing to reveal. Such prejudices provoke reactions that may not be altogether friendly. By contrast, when the sound of a familiar accent reaches our ears we know immediately that the speaker is 'one of us'. This may generate a natural sympathy and open the door to dialogue. Is it the same for two chaffinches who meet far from home? Do they say to themselves, 'Ah, you're from where I'm from'? In the end, we are not much more sophisticated than chaffinches.

20.
MAKING LOVE

The rational penguin or the passionate duck?

IS THERE ANY TRUTH TO the notion of 'animal passion'? It is based on the assumption that, rather than making love, animals do the deed without inhibition or sentiment. Birds are just one species on the long list of supposedly inferior creatures for whom the sexual act is presumed to be something brutish and violent. In the case of mallard ducks, this is quite true.

In the middle of winter, very early on in the mating season, male mallards don their courtship plumage and begin to mill around the females. But male ducks commonly outnumber females, and it isn't rare to see three, four, five or even six male mallards courting one poor female, who is usually trying desperately to escape. To no avail. She is quickly surrounded and one male climbs on top of her, then another climbs onto the first and a third on top of the second . . . Eventually, the unfortunate female may even perish under the continued assault of the males.

Not all birds behave like this, however; far from it. While a bird won't actually serenade its beloved on the mandolin, some come pretty close. Take the example of terns, also known as sea swallows because of their slim wings, long forked tails and agile flight. When the courting season comes around, the male doesn't force himself boorishly on a female, but makes it his business to seduce her, with patience and gifts. By way of encouragement, he brings offerings of little fish, which he presents to her at the nest, presumably demonstrating that he is good at fishing and will provide food for their future offspring. These gifts work their magic and pave the way for a very gentle mating process, but they also reinforce the ties between the couple, who will bring up their young together until they are old enough to live independently.

The gentoo penguin channels his passion in a similar way. The male presents his partner with an endless succession of stones. He is not very agile on land, but nonetheless waddles back and forth between the beach and the place where the female will lay her eggs, collecting stones one by one and carrying them carefully in his beak. He places each one at the female's feet before going off to find another. Then, when there are enough stones, he arranges them (lovingly, shall we say) in a little circle around the place where the eggs will be laid. This is a useless construction because the eggs are in fact laid unceremoniously straight on to the ground, but despite this the female likes the stones and defends them fiercely if a neighbouring bird tries to scrounge any.

Are some birds more rational, more sensible, than others? Our judgement of this is inevitably coloured by our human sense of what is and isn't reasonable, but it does seem clear that for the more 'rational', less 'passionate' species, mating is a less tempestuous affair. By eschewing the sexual opportunism of the mallard and other species, terns and gentoos, for example, are more successful at continuing their species, because both partners are involved in looking after the chicks, unlike the male mallard who deserts his female and future young as soon as mating is over. The more 'rational' birds seem to have a greater chance of raising their young unharmed than those that have a brief encounter after which the female is left alone to look after the chicks. Perhaps this is the same for us. We humans could no doubt see some kind of 'moral' in all this, but birds just do what they do in order to perpetuate the species.

21.
WHAT CAN BEAUTY TELL US?

Not a feather out of place

OUR HUMAN EYES DELIGHT in the colours of butterflies, tropical fish and birds. For us, these creatures are incarnations of perfect beauty. Birds in particular combine an array of charms: naturally elegant flight; feathers that can be long and ornamental, embellishing the bird's shape (such as in the case of the egret's filoplumes, which are fine, hair-like feathers); and a seemingly infinite kaleidoscope of colours. Their song is the finishing touch, setting the whole multicoloured spectacle to music. At least, this is true of those species that possess impressive plumage. There are many that don't, as we will see.

For the birds themselves, coloured feathers have an important behavioural function: allowing the birds to communicate with each other and engage in successful courtship. Still, the idea of beauty 'for beauty's sake' should not be completely discounted.

For humans, ideal beauty tends to take a female form, but the opposite is true of many bird species, where the males sport the brightest colours. The females' more discreet plumage allows them to melt into the background easily when they are sitting on their nests or caring for their young. Dull plumage is a good protection against predators. It really is a matter of life and death: the brightly coloured males become easy prey in spring when their feathers are at their most impressive. There are some exceptions to this, in those species for which colourful plumage doubles as camouflage. The golden oriole, with its yellow and black plumage, is a good example. In flight, the oriole's vivid colours make it difficult to miss, but when it alights in a tree (and it is a forest-dwelling bird), it becomes completely invisible simply because its combination of yellow and black perfectly mimics dappled forest light.

For most males, however, all this finery carries a certain risk, but it is necessary if they are to attract a female. She will often choose the most flamboyant bird as her mate, taking his exuberance as proof of his health and vigour: the more he shows off, the better his genes are likely to be.

Incidentally, male and female birds of species that share the task of bringing up their young often don't exhibit sexual dimorphism – in other words, they look the same as one another. This makes it very difficult to tell the difference between male and female seagulls or crows, for example. Sexual dimorphism is not always striking among humans either, which is why we mark our genders by artificial means: hairstyles, clothes, typically masculine or feminine gestures . . . different cars. The pressure to categorise oneself is very strong and those who transgress the norms may be judged harshly.

This question aside, beauty is undeniably central to seduction among humans just as it is among birds. We may even cheat a little and use a touch of artifice to improve our physique. But perhaps seduction always involves a little trickery. Think of the myth of Jupiter transforming himself into a swan in order to seduce Leda and take advantage of her gullibility: seduction and sincerity don't always go hand in hand, and artifice can be taken to such lengths that it becomes complete transformation. This is also true of some male birds who actually 'metamorphose' in spring, developing splendid plumage that they shed as soon as mating is complete. Cruel deception! Rather like the behaviour of people who deploy all of their beauty and finery for the purpose of seduction, only to completely neglect such things once they feel that their partner is won over.

If we delve a little deeper into this subject we discover that a bird's fine plumage conceals the lack of something else: the ability to sing. Although there are some exceptions, the males of the most sumptuously attired species only produce very mediocre songs. Vocalisation, like plumage, can be a key element in the seduction process and, by contrast, the best singers are often those species with the most modest plumage: the nightingale clad all in brown, the blackbird or the song thrush with its scattering of pale beige spots.

They are modest virtuosos. It would seem that evolution has given birds only two options: feathers or song. Some, like crows, have neither. But at least they're clever, and you can't have everything.

A recent study came up with an interesting finding: in one particular species of finch – the siskin – the male has yellow, green and black plumage and the females select the most brightly coloured, and therefore most beautiful, males because they are also the cleverest. How can this be? The birds are not able to naturally produce certain pigments themselves (yellow, red and orange) but can absorb them through their food, at which point the pigments colour their plumage. Thus, those males that find the most food have the most vivid plumage, and this fact doesn't escape the females: the most brightly coloured male is the best at finding food for himself – and therefore also for his young! Here 'beauty' is an important element in the process of selection, with the male siskin signalling his suitability via his plumage. A well-dressed man or woman will attract attention in a similar way. Seduction is largely a matter of using our attributes intelligently to show ourselves in the best possible light. Whether there is a process of reproductive selection going on behind the scenes, the reader can decide for themselves!

22.
LEARNING TO DIE,
LEARNING TO LIVE

Swallows hide away to die

BIRDS HIDE AWAY TO DIE, so they say. And it's true. Have you ever seen a dead swallow, except for one that's been hit by a car or flown into a pane of glass? Do you ever come across the carcasses of birds? No. Because the ill or weak bird either gets caught – and eaten – by a predator or has the time to go and conceal itself somewhere before it breathes its last.

Birds don't have long-term illnesses or very long lives. As soon as a bird is no longer in good health, nature goes about ending its life. Is this cruel? Or is it we humans who are cruel when we prolong life beyond its natural course, making old or terminally ill people endure weeks and weeks of suffering? Nature does not allow pain to last very long. Death throes are always brief, and there is no such thing as a slow decline, whether physical or mental. At its heart, in the world of birds, nature has chosen well.

Montaigne said that to study philosophy is to learn to die. Perhaps philosophy can help us prepare for death, but is that really possible? All philosophies and religions say the same thing: the best way to reconcile ourselves to the inevitable end that awaits us and those we love is to live life fully, in the present moment. We can appreciate and savour the gift of the present moment: a ray of sunshine, a juicy peach, an unexpected smile, a collared dove pecking about in the garden, a crested tit performing acrobatics on a branch . . . But does the little crested tit need to reconcile itself to death? No, of course not. Because it already makes the most of every moment, appreciates every seed and each ray of sunshine. It doesn't need us to teach it this truth, it doesn't need to philosophise – it is already living its life to the full. Are birds full of wisdom, then? The crested tit doesn't contemplate its own existence, doesn't plan, doesn't put things off, doesn't pretend that things will be better in the future. It simply lives.

How many times a day do we hear people claiming that their lives will be better 'one day': when they have found love, when they have gone on holiday or retired, when they have found a new job or got a raise? But 'one day' often means never. Of course, we need to dream and some changes are useful. But life is here, now. Who knows if we will still be alive this evening? Who knows what death has in store for us? And who knows what it has in store for those we love most? Why not be a little more like birds and live in the intensity of the present and thus avoid dying with a heart full of regret?

Birds don't think about their death. They have the good luck not to be able to conceive of it and intellectualise it as we do, but that doesn't prevent them from being aware of their own fragility and finite nature, doing everything they can to survive and being wary of all predators. And, in the end, it's true that there's no point thinking about death or worrying about it when we aren't in any actual danger, because worrying about it doesn't change anything.

Death and life are one and the same thing, one does not come without the other: this is the immutable law governing us all, people, animals, plants. Our existence is full of little deaths, mournings, ruptures, beginnings and rebirths. And, in any case, biologically speaking nothing ever truly dies: at the moment of our death our atoms do not disappear; they are recycled, and some will become part of a worm or a flower, which may be eaten by a bird, and so on. Eastern philosophy is founded on the idea of cycles, something that the West, with its more linear vision, is inclined to forget. But nature and birds remind us, and rightly so.

Perhaps we don't need to learn to die. Perhaps we need to learn to live.

Conclusion: adapt or perish?

IN OUR CHANGING WORLD, threatened by climate change and the destruction of natural habitats, many bird species are disappearing. And what about us? Will we survive in the artificial world that we are creating?

How do species adapt to their environments over time? This is one of the big evolutionary questions and it is key to the appearance and disappearance of all species, despite any notions we humans may have about escaping extinction thanks to our extraordinary intelligence, or perhaps through divine intervention. Many of us don't feel worried. And yet ...

It was Darwin who first shed light on the evolutionary process that has caused millions of species to appear, flourish and disappear since the beginning of life on earth. A species transforms into another species, then another, and another and so on. The briefest of glances back at our history tells us that evolution moves slowly, very slowly. Birds arrived on the scene around 150 million years ago – and they didn't appear overnight. They have their origins in the group of dinosaurs called theropods and are thought to be their only living descendants. The progression from velociraptor to archaeopteryx to goldfinch – to give a rather potted history – took millions of years. Soft feathers, the power to fly long distances and the ability to sing didn't all appear by magic. There were periods when the number and diversity of species exploded, often followed by mass extinctions in ongoing cycles. During these periods of profusion and decline, one species might give rise to a new one, or disappear without trace. There were tens of thousands of failed experiments that led to evolutionary dead ends. It was in this crucible of life that birds, mammals and indeed all living creatures came into being.

Humans are no exception, having their origins in a line of primates that gave rise, eventually, to the genus *Homo*, of which we are the latest avatar . . . for now, at least. Until recently, *Homo sapiens* (the self-declared 'wise' branch of the Homo genus) were evolving in more or less the same way as other species – that is, slowly. We adapted to our environment and, with the help of our large brains, gradually improved our daily lives, health and life expectancy. Over time we conferred some of these benefits to other species that shared our lives (which is how domestic animals such as dogs, cats and horses came about). But all other living creatures, including birds, are still governed by the basic evolutionary process.

All was for the best in the best of all possible worlds

until man stepped on the accelerator. By exploiting the planet for our own ends and thus completely upsetting its natural balance, we have caused serious problems not only within the earth's ecosystem but also, in the last few decades, for its climate. The long periods of time needed for species to evolve have, in their turn, been disrupted. The human race now lives in a state of immediacy; we are constantly seeking to go faster, produce more, build more and modify our behaviour more quickly.

But life on earth wasn't prepared for all this. Human pressure on the natural world has forced many species to adapt (as quickly as they are able to) or perish. This is why scientists are increasingly talking about a sixth mass extinction, bringing about an enormous reduction in biodiversity that will in turn cause our ecosystems to become extremely fragile. This isn't the first mass extinction, people argue, there have already been five, all before people appeared on earth. And each time life began again. True, but this one is different; this one would be caused by human activity. Previous extinctions occurred over a relatively long period – and it took even longer for life to re-establish itself – but the one we now face will be devastatingly sudden and leave little hope of a speedy recovery.

We need to open our eyes and our ears. Anyone

who takes even a passing interest in the little flying, singing creatures around us will have noticed that the number of skylarks, swallows and many other birds has fallen sharply in a very short space of time – barely a few decades, which is nothing in evolutionary terms. A few species have succeeded in adapting very rapidly, but for most, adapting to an ecosystem in permanent upheaval is difficult or impossible. This is why it comes as no surprise that twenty-five per cent of the earth's ten thousand bird species is threatened with extinction by the end of the twenty-first century. Tens of thousands of animal and plant species will also be lost.

It is the most specialised bird species and those living in ecosystems that only cover very small areas that are the most vulnerable and likely to disappear first. But they're not the only ones we should worry about. There may come a time when even the most adaptable species will be caught in the trap, and humankind may be among these, the final victim of an evolutionary crash caused by our misguided use of the accelerator. Is this what we want? A hostile environment? A world without birds, in which we will have to explain to our children why we destroyed the swallows? If we let this happen then we are clipping our own wings. It may please us to believe that we are all-powerful, above other species, that we can use nature to our own ends, but this is an illusion. We are at a crossroads. Our destiny is in our hands, like a chaffinch almost crushed in our grip, its heart beating fast, desperate to fly away. We can decide whether to open our hands and let the bird fly free, or smother it completely. Perhaps the last lesson to learn is the most obvious: the day we decide to protect birds will be the day we decide to protect ourselves.

About the Authors

Philippe J. Dubois is an ornithologist and writer. His passion for birds began in childhood and his bird-watching has taken him many places around the world.

Élise Rousseau is a philosopher and journalist. She has written numerous books about nature and animals and also works for the protection of the environment.